Table of Contents

1. Introduction

Fresh horticultural products like fruits, vegetables and flowers are highly perishable and hence appropriate product-specific packaging assumes great significance in preventing their rapid deterioration during storage and transport. Packaging is an important step in the post harvest management of any horticultural product and major objective of packaging of any horticultural commodity should be the prevention of physical damage and microbial contamination during its storage, transport and distribution while taking appropriate post harvest measures to ensure the maintenance of its 'fresh-like' quality as long as possible. Shelf life extension of a fresh produce may successfully be achieved by adopting proper packaging solutions for preventing dehydration or dampness of the product and its nutritional losses while keeping the product in optimal storage conditions.

Traditional packaging of horticultural produce in gunny bags, wooden crates, bamboo baskets, cartons, bulk bins, and palletized containers has now replaced with modern packaging solutions such as modified atmosphere packaging, modified interactive packaging and controlled atmosphere packaging solutions. Currently, there are thousands of different types of packaging solutions available in global markets and the number continues to increase as the packaging industry regularly innovates and introduces new packaging concepts with a wide range of packaging sizes to meet the diverse requirements of producers, processors, buyers, wholesalers, retailers, shippers and consumers.

2. Current Packaging Practices

Traditional packaging systems are still popularly practiced in third world countries including India and China. This is mainly because of the economic viability, biodegradability and cheap availability of traditional packaging materials like corrugated fiber board (CFBs), jute bags, paper and pulp. However, developed nations have already started adopting most modern packaging technologies like modified atmosphere packaging, modified interactive packaging, active and intelligent packaging etc in the field of fresh produce packaging. An illustration of current packaging systems for fresh fruits and vegetables as practiced in India and USA is shown in Table 1 and 2 respectively.

Table 1: Current Packaging Practices in India

Product	Packaging Solutions
Oranges	Molded pulp trays in CFBs: Each tray contains about 20/25/30 oranges and four to five such trays are placed in a corrugated fiberboard box
Custard Apples	3-Ply CFB boxes of RSC type or EPS (Expanded Polystyrene) boxes of 3kg capacity
Lychees	Eight plastic punnets (each of 250 grams capacity) in a CFB box of 4 Kg capacity or 3-Ply CFB boxes of RSC type of capacity 2 kg
Green chillies	3-Ply or 5-Ply CFB boxes of capacity 5kg to 10 kg
Yams	Gunny bags of 10kg to 15kg capacity
Curry leaves	Plastic pouches of 30gm capacity in 3-Ply CFB box
Potatoes	Jute bags of 25 Kg capacity
Gourds	3-Ply or 5-Ply CFB boxes of capacity 5kg to 10 kg
Brinjal	3-Ply or 5-Ply CFB boxes of capacity 5kg to 10 kg

Source: (IIP, 2005)

Table 2: Current Packaging Practices in USA

Packaging for Perishable Products including Fruits and Vegetables	Packaging Solutions Used are CFBs (most popular) and Plastic and Wood Boxes (limited usage)

Field Packing	Hand packing in corrugated fiberboard boxes
Packing at Pack Houses	Mechanical Packing
Shipping Practices	Products are unitized and shipped on pallets

Source: (Kader, 2002)

3. Packaging Materials Available Today

Traditionally, paper, pulp and jute are being used to pack various horticultural commodities. These packaging materials are not only cheap but biodegradable also. Jute bags and carton boards are one of the oldest traditional types of packaging solutions and are still in use today. Jute is the most cost effective, high tensile vegetable fibers and therefore jute bags are used as the most eco-friendly packaging option in the bulk packaging sector while paper bags are extensively used in the retail sector. However, in modern times, the use of paper, pulp and jute as packaging materials is getting replaced with eco-friendly polymer-based packaging solutions.

3.1   Paper and Wood -based Packaging Materials

Paper and wood-based packaging materials are very popular and still in high demand among fresh produce suppliers and retailers. There are mainly two factors to be taken into consideration while using paper and wood-based packaging: appearance properties and performance properties. Appearance properties are color, surface smoothness, surface structure, gloss, opacity, printability, varnishability, surface strength, ink and varnish absorption and drying, surface tension, rub resistance, and surface cleanliness. Properties that contribute to the performance are weight, thickness, moisture content, tensile strength, tearing resistance, burst resistance, stiffness, compression strength, creasability and foldability, ply

bond (interlayer) strength, flatness and dimensional stability, porosity, water absorbency, gluability, adhesion, sealability, odor neutrality and product safety. Major sources that can be used for the manufacturing of paper and wood-based packaging options are wood pulp, fresh fiber, forest products, recovered or recycled fiber, and pressed paper. Major paper-based packaging options available in the market include greaseproof paper, tissue paper, label paper, vegetable parchment paper, bag papers, impregnated papers, laminated papers, SBB (solid bleached board), SUB (solid unbleached board), FBB (Folding Box Board); and WLC (white lined chip board). (Kirwan, 2010). Major packaging solutions available in corrugated fiber board (CFB) include boxes with molded pulp trays and boxes with slotted partitions. Smurfit Kappa group, a global leader in providing paper-based packaging solutions has introduced a variety of packaging options in the market. In most European countries Smurfit Kappa is a leading supplier of boxes and trays for fruits, vegetables, perishables and flowers.

3.2. Polymer-based Packaging Materials

Use of various types of polymers and polymer-based materials like polystyrene, polypropylene, polyvinyl butyral, polyvinyl alcohol, ethylene vinyl acetate, and polyethylene is gaining momentum and popularity due to multitudes of advantages these packaging materials offer to both the wholesale and retail sectors.

4

3.2.1. Polyethylene (PE)

PE is used for the production of shrink films; flow pack films; top seal films; cover sheets for trays; single film bags, and tubular netting. Knitted tubular netting made from PE is an attractive type of packaging for potatoes, vegetables and fruits. It is strong, economical in use and has a large number of different processing applications. High Density Poly Ethylene and Low Density Poly Ethylene are two major types of PE used for the production of various sizes of sealable bags, PE films and tubular nets.

3.2.1.1. High Density Poly Ethylene (HDPE)

HDPE has good barrier properties and therefore mainly used for making Stretch or Cling Films; and Raschel Bags.

3.2.1.2. Low Density Poly Ethylene (LDPE)

LDPE has Low Water Vapor Transmission Rate (WVTR) but High Oxygen Transmission Rate (OTR). Therefore LDPE is used as a base material for modified atmosphere packaging (MAP) films. LDPE is abundantly used for the production of lidding, base webs, and trays. LDPE is used for making monofilament net films (knitted/woven) and tubular netting (knitted/woven) also.

3.2.2. Vinyl Polymers

Major vinyl polymers that are used in the packaging industry are Poly Vinyl Chloride (PVC) and Ethylene Vinyl Acetate (EVA).

3.2.2.1. PVC (Poly Vinyl Chloride)

PVC is used for producing top quality stretch films, which are produced by cast extrusion. This method of extrusion ensures low tolerance levels, consistent quality, providing excellent strength and stretch. This unique strength of the films (due to low manufacturing tolerance levels) can enable food processors to reduce the thickness and therefore the costs of the films they use. There are single layer PVC films and multilayered PVC films. Single layer PVC films, although very tough, are in fact single layer and therefore fully recyclable with relatively low waste disposal costs. PVC films are good gas barriers but moderate oxygen barriers. PVC films are extensively used for the production of thermoformed trays.

3.2.2.2. uPVC (Unplasticised Polyvinyl Chloride)

uPVC films (unplasticised polyvinyl chloride film) are made from PVA/PVOH (poly vinyl alcohol), a water-soluble synthetic polymer. This is a highly cost-effective thermoformed packaging material and is available in a wide variety of gauges and finishes. Major properties of uPVC films include excellent film forming, emulsifying, and adhesive properties. It is resistant to oil, grease and solvent.

3.2.2.3. Polyvinylidene chloride (PVdC)

Polyvinylidene chloride, a polymer derived from vinylidene chloride is applied as a water-based coating to enhance the barrier properties of other plastic films such as BOPP and PET.

3.2.2.4. PVOH (Poly Vinyl Alcohol)

PVOH is another polymer-based packaging material available in the market. Major properties of PVOH include high tensile strength, flexibility, as well as high oxygen and aroma barrier. It has a melting point of 230°C and 180-190°C for the fully hydrolyzed and partially hydrolyzed grades. It decomposes rapidly above 200°C and acts as a carbon dioxide barrier in polyethylene terephthalate (PET) bottles. PVOH may be used as a biodegradable plastic baking sheet and also as a water-soluble film useful for packaging. PVOH is extensively used for making water-soluble packaging films and biodegradable baking sheets.

3.2.2.5. EVA (Ethylene Vinyl Acetate)

EVA is a copolymer of ethylene and vinyl acetate). Major properties of EVA include good clarity and gloss, good barrier properties (High WVTR& OTR), low-temperature toughness, stress-crack resistance, hot-melt adhesive and heat sealing properties and resistance to UV radiation. EVA emits little or no odor and is competitive with rubber and vinyl products in many electrical applications. It is used as a shock absorber and popularly known as 'expanded rubber' or 'foam rubber'. EVA is extensively used for the production of lids, base webs and trays.

3.2.2.6. EVOH (Ethylene Vinyl Alcohol)

Ethylene Vinyl Alcohol, a copolymer of ethylene and vinyl alcohol is commonly used in food packaging applications primarily as an oxygen barrier for shelf life extension. EVOH is defined on the basis of ethylene content: lower ethylene content grades have higher barrier properties and higher ethylene content grades have lower barrier properties.

3.2.3. Polystyrene (PS)

Polystyrene is a highly transparent polymer used for the production of various types of packing materials. Polystyrene-based packaging materials that are popular in the packaging industry are OPS (Oriented Polystyrene); EPS (Expanded Polystyrene); HIPS (High Impact Polystyrene) and XPS (Extruded Polystyrene).

3.2.3.1. OPS (Oriented Polystyrene)

Oriented Polystyrene is highly flexible and is used for the production of flexible packaging options such as flexible pouches. It can be used as a molded packing material. It is mainly used for the production of Oriented Polystyrene Films and Thermoformed Trays.

3.2.3.2. EPS (Expanded Polystyrene)

EPS is normal polystyrene expanded into foam through the use of heat. It is a rigid cellular plastic, which can be presented in several ways, adapting to different uses. This material is used for fruit and vegetables crates, and other agricultural and gardening products. The EPS package is the best packaging solution for perfect food preservation as it provides best standards of sanitary quality, thermal insulation and chemical resistance. It also provides resistance and protection against moisture and shocks. It is resistant to compression and has good display effects which work in favor of visual merchandising. It is extraordinarily light weight and has ease of conformity. EPS packages can easily be customized through several packaging techniques. Several researches carried out in various fruits and vegetables have now confirmed the fact that the freshness of fruits and vegetables is best if preserved in EPS packages. This is basically due to the effects of thermal and moisture protection. When longer the keeping period, this effect is clearer, comparatively to other packaging materials. EPS packages are made of 100% recyclable material. Several analyses of EPS show that it is highly eco-friendly. Moreover, EPS manufacturers are members of CICLOPLAST, an organization that guarantees compliance with the Packaging and Residues Law - Ley 11/1997. One of the leading manufacturers of EPS packages is Styropack Inc, Denmark. They supply EPS boxes for fruits and vegetables. Delicate fruits and vegetables need to be protected in stable and water-resistant packaging during transportation. For this purpose EPS boxes are very well suited, as the material is shock-absorbing, thermal insulating and light-weight. It is very easy to handle also. Another global player in the field of EPS packaging is Associated Packaging Inc. They manufacture and supply a wide range of finely finished EPS produce trays profile of which ensures that the consumer has maximum product visibility.

3.2.3.3. Extruded Polystyrene (XPS)

XPS is different from expanded polystyrene (EPS) and is commonly known by the trade name Styrofoam. It has low thermal conductivity and makes a more-uniform substitute for corrugated cardboard.

3.2.3.4. High Impact Polystyrene (HIPS)

High impact polystyrene is used for the production of HIPS (high impact polystyrene) punnets and thermoformed trays. A major player in providing HIPS packaging solutions is

Sealed Air Corporation. Its 'Cryovac Barrier Trays' are made from HIPS and come in a wide range of sizes, depths and colors. These HIPS trays have excellent barrier performance characteristics to maximize product shelf life.

3.2.4. Polyethylene terephthalate (PET or PETE)/Polyester

PET is an effective thermoforming material with many finishes and a good range of temperature performances. Sharp Interpack, a global player in providing PET packaging solutions, is a specialist producer of CPET (crystallized PET) material for dual-ovenable packaging solutions and APET (amorphous PET) for high clarity food packaging. PET is used for the production of multilayered laminates through lamination process. Associated Packaging Inc, another leader in PET packaging solutions has introduced TAIRILIN polyester films (PET) in the market. It is used for the packaging of microwaveable and ovenable food products. TAIRILIN is a biaxially oriented polyester film with excellent tensile strength; excellent mechanical characteristics; chemoresistance; heat resistance; excellent transparency; glossy surface; good dimensional and chemical stability; excellent electrical characteristics and excellent heat sealability

3.2.4.1. CPET: Crystallized PET

CPET, a thermoplastic polymer used for manufacturing synthetic fibers; food containers and thermoforming applications. CPET is the most common packaging material used in dual-ovenable packaging applications. CPET is also used for the production of bottles; trays and punnets. Its service temperature ranges from -40°C to +220°C and hence suitable for freezers, microwaves, conventional ovens and industrial ovens.

3.2.4.2. APET: Amorphous PET

APET is best suitable for the packaging of cold and ambient temperature food products. It serves as excellent barrier for MAP systems. APET has high clarity and hence used for high clarity food packaging. A standard product of APET is freezable to -18°C and its service temperature ranges from -40°C to $+65^{\circ}$C. It is used for the production of lidding, thermoformed trays, bottles; and punnets.

3.2.5. Polypropylene (PP)

Polypropylene is a stiff and tough thermoplastic polymer that can be made either transparent or opaque. It is very economical to use. It has a melting point of 160°C and hence suitable for making food containers that are suitable for the dishwasher. It is strong enough to withstand industrial hot filling process. Polypropylene is mainly used for the production of trays, lenobags and punnets. It is widely used in manufacturing microwaveable packaging materials as well as modified atmosphere packaging materials. Special formulations of polypropylene are available for freezer applications. Various types of Polypropylene (PP) packaging films available in the market are BOPP (Biaxially Oriented Polypropylene film); CPP (Cast Polypropylene film); MOPP (Mono-Oriented Polypropylene film) and NWPP (Non-Woven Polypropylene film).

3.2.5.1. Amcor Seal-Plus, Polypropylene packaging from Amcor Flexibles

Amcor Flexibles, a global supplier of packaging solutions for horticultural sector has launched a concept called Amcor Seal-Plus in polypropylene packaging. This innovative packaging technology consists of a unique OPP film with an improved seal performance called 'SealPlus'. The main benefits of SealPlus over standard OPP film are its outstanding hot tack performance (this allows for the potential to increase machine speeds for heavier/dense fast dropping products), up to 1Kg recommended maximum weight (compared to 750g for standard OPP), its superior seal integrity and more consistent gas levels (therefore more consistent product).

3.2.5. Polyamides (PA)

It is a naturally-occurring polymer that contains monomers of amides held together by strong peptide bonds. For artificial synthesis of PA, polymerization or solid-phase synthesis is used. Artificially-made PA is stronger and more durable making it suitable for a wide range of food packaging applications.

3.3. Ecofriendly Packaging Materials

Current trend in the fresh produce packaging sector is characterized by the high reliance on ecofriendly packaging materials such as biopackaging materials from corn starch and potato starch and PLA packaging materials from fermented agricultural products.

3.3.1. Biopackaging Materials

Biopackaging materials are manufactured from renewable raw materials like starch from maize or grains; potato starch and fossil fuels. Biopackaging is now widely practiced in ecofriendly manufacturing facilities. Asia-Pacific based Convex Plastics is a leading supplier of biopackaging material in the world.

3.3.1.1. GreenSACK from Convex Plastics

GreenSACK is a soft, opaque biodegradable packaging material introduced by Convex Plastics. It is made from corn starch and has a good strength as that of PE. It can easily be disposed of by using it as a composting material. It is degraded within 5 -6 weeks and is safe to the environment. It is mainly used for making agricultural films, magazine covers, green waste bags and carry bags.

3.3.2. PLA Packaging Materials

PLA films, an ecofriendly packaging solution are increasingly becoming a viable packaging option for fresh produce sector. PLA is bio-based as it is derived from the fermentation of agricultural products and also biodegradable. PLA is popular as a sustainable alternative to petrochemical-derived polymers. Since PLA does not leave waste and can be composted in industrial facilities itself, it becomes a preferred packaging solution for the facilities where waste management becomes a herculean task. Nevertheless, PLA has certain disadvantages also. It is highly brittle and less durable as compared to petroleum-derived polymers. However recent researches have shown that this flaw can be rectified by using certain additives to incorporate into the PLA films during the manufacturing process. Asia-Pacific based Convex Plastics is a major supplier of PLA packaging solutions in the world.

3.3.2.1. NatureWorks from Convex Plastics

NatureWorks is a PLA packaging material introduced by Convex Plastics. It is made from fermented corn and is easily degradable into compost within a span of 4 -6 weeks. Its best use is for the production of lidding films.

4. Modern Innovations in Packaging Materials

Major modern innovations in the field of fresh produce packaging are modified atmosphere films, modified interactive films and packing films for active and intelligent packing.

4.1. MAP Films

Post harvest quality retention is a critical issue in the perishable produce market. Most of the available techniques tend to affect the natural, 'fresh-like' quality of the fresh produce and hence a produce-friendly packaging material is necessary for fresh produce packaging. MAP films, also called 'modified atmosphere films' are delicate, produce friendly and are the most suitable packing materials that are available in the current global market. MAP films are made using polymers having high WVTR (water vapor transmission rate) and low OTR (oxygen transmission rate) so that these permeability properties of the polymer films can advantageously be used for creating a modified atmosphere inside a package. Packing material having high WVTR releases excess moisture present inside the package into the surrounding atmosphere while its low OTR properties can be manipulated to maintain optimum gas composition inside the package. Major polymer-based packing films that can be used as MAP films are EVOH, PVC, PA (polyamides), PE, PP, PET, PS (polystyrene) and PvdC. Selection of MAP film for a fresh produce should be based on its gas and vapor barrier properties, optical properties, antifogging properties, mechanical properties, and heat sealing properties. (Richard Coles, 2003)

Modified atmosphere films could be manufactured using laser induced micro - perforated technology. First of all, a "permeable film" is created which when hermetically sealed becomes conducive for fresh fruits and vegetables to survive for longer periods under

14

a rigid cool chain environment. Product quality is maintained and product shelf life is prolonged due to MAP films and this is a combined effect of Modified Atmosphere (high CO_2 and low O_2); Modified Humidity (90-95 % RH) and condensation control (removal of excess moisture).

It is beyond doubt that product-specific factors such as respiration, transpiration, and ethylene sensitivity play major roles in determining product quality and thereby product shelf life in horticultural products. Therefore the type of MAP packing material required by the products also differs from product to product depending on the product-specific physiological factors. Hence it is important to know the product-specific respiration rate, transpiration rate, moisture content and other values so that MAP films with different properties can be manufactured to suit specific produce requirements. The point here is, it is simply not possible to use the same MAP film for all produce. Same is applicable for temperature regulation also. If the correct temperature is not used for the MAP, produce respiration will rapidly increase causing high levels of water vapors. This creates post -plant stress and as a result ethylene gas levels also start rising thus creating the wrong environment for the produce. So while adopting MAP technology, it is highly recommended to choose right MAP film and to use correct temperature. A list of various packing films suitable for MAP is given in Table 3.

Table 3: Packaging materials for MAP

Packaging material	Properties
Polyethylene	

LDPE	Low WVTR, High OTR
LLDPE	Good impact ,tear ,tensile puncture
HDPE	Superior barrier properties than above
Polypropylene	
OPP	Low WVTR, Low OTR
COPP	Low WVTR, Low OTR
Vinyl Polymers	
EVA	High WVTR& OTR (higher than LDPE)
PVC	Good gas barrier, moderate O2 barrier
PVdC	Outstanding Barrier properties
EVOH	Very high gas barrier, moisture sensitive
Polystyrene: HIPS	High tensile, low barrier prop.
Polyamide: Nylon -6	Good barrier
Polyesters: PET	High Clarity

4.2. MIP Films

Packing quality of a packing material can be enhanced by incorporating additives and minerals into it while it is being manufactured. MIP films, also called 'modified interactive packaging films' are usually made from impregnated mineral rich polymer-based packing materials.

4.3. Active and Intelligent Packing Films

Being the most modern innovation in the field of fresh produce packaging, active and intelligent packaging technology makes use of various modifications of MAP films, for example non-toxic insect-repellant MAP films, MAP films that prevent the growth of food-borne pathogens, and high vapor-permeable MAP films, in order to ensure food safety and quality maintenance. (Wilson Charles L. Ph.D, 2007). Modified MAP films are often used in combination with additives having antimicrobial properties. Materials having gas absorbing and releasing properties are also used in ceratin cases to enahnce the performance of intelligent packaging technology.

5. Manufacturing of Polymer-based Packing Materials

Various manufacturing technologies are used for the production of polymer-based packing materials. Thermoforming, vacuum forming, pressure forming, twin-sheet forming, drape forming, free blowing, simple sheet bending, lamination, extrusion, co-extrusion and conversion technologies are used for making various types of packing materials.

5.1. Thermoforming

Thermoforming is a process of manufacturing of thermoplastic films. The film is heated between infrared, natural gas, and other heaters to its forming temperature and then it is stretched over or into a temperature-controlled, single-surface mold to manufacture thermoplastic sheets. Thermoplastic sheets are extensively used in the food packaging industry for the production of thermoformed trays.

5.2. Lamination

Lamination is a process by which two or more packing films are pressed together to produce single layered or multilayered laminates. APET/PE lamination is used for manufacturing  MAP (modified atmosphere packaging) films. Water-resistant laminate technology is used for producing water-resistant laminates. Oil-resistant laminates are also produced using the same technology. AMCOR Inc, a multinational supplier of packaging solutions has already introduced a series of laminates suitable for fresh produce packaging under the trade name, **Amcor Panorama.** Amcor Panorama, a paper strip laminated to a variety of substrates, combines the natural feel of paper with the latest flexible technology for the product visibility.

5.3. Extrusion

Extrusion is a manufacturing process of polymer-based packing material where heat is applied to melt and form the packing material (normally single layered) of desired dimensions. Co-Extrusion is another type of extrusion manufacturing technology employed in manufacturing multi-layered packing materials.

5.4. Conversion

Conversion is a process by which one form of plastic material is converted to another form using automated converting systems.

6. Selection of Suitable Packing Material

While selecting a suitable packing material for fruits and vegetables, both packing material properties and product-specific properties must be taken in to consideration. Major packing material properties that are to be considered are its ecofriendly characteristics, permeability properties and sealing characteristics while product-specific characteristics are its physiological requirements and shelf life properties. In addition to this, a packaging material must add value to the product and it should be economically viable.

6.1. Packing Material Properties

Major packing material properties that are to be considered for fruit and vegetable packaging are recyclability or reusability; biodegradability; permeability; ease of disposal; sealability; thermoformability; tension puncture; transparency; light weight and suitability to the product to be packed.

6.1.1. Biodegradability of the Packing Material

Packing material that is to be used for fruit or vegetable packaging must be ecofriendly and safe to the use of humans. It should be reusable (recyclable), biodegradable and easy to dispose of. It should have good waste management properties. Some of ecofriendly packing materials include PET punnets with lid, EPS tray stretch wrapped, leno/raschel/net bags and molded EPS box with lid. Environmentally safe packing materials are gaining popularity nowadays and therefore packaging materials that are recyclable or biodegradable, or both are highly recommended for fresh produce packaging.

6.1.2. Permeability Properties of the Packing Material

Permeability properties of the packing material in terms its WVTR (Water Vapor Transmission Rate) and OTR (Oxygen Transmission Rate) must be taken into consideration while selecting it for fruit or vegetable packing. Tension puncture properties of the material should be in tandem with the physical properties of the product to be packed.

6.1.3. Sealing Characteristics of the Packing Material

Sealing characteristics of the packing material should be measured and tested at film selection and again at package converting and product fill stages. Major factors to be considered include film characteristics such as heat sealability, thermoformability, strength and endurance.

6.2. *Product-specific Properties*

A fresh fruit or vegetable continues to respire and metabolically active even after its harvest. Physiological characteristics of fruits and vegetables such as respiration, transpiration, moisture content, ethylene sensitivity, degree of microbial contamination, presence of mechanical injuries and physiological maturity play an important role in determining a suitable packaging option for them.

6.2.1 Product Physiological Requirements

Packing material selection should be based on the produce physiological requirements such as oxygen and carbon dioxide requirements based on product respiration rate; water requirements based on product transpiration rate; and cold storage requirement for maintaining optimum temperature for storage and transport. Packing material selection should also be based on the nature/type of the produce to be packed; its pre-harvest conditions and post harvest conditions.

6.2.2   Shelf life Properties of the Product

Desired shelf life of the product should be calculated and accordingly packing material should be custom-engineered for maximizing the shelf life of each product based on its product-specific characteristics.

7. Packaging Designs

Major parameters that should be considered while designing a package for a fruit or a vegetable are packing material properties, product requirements, and consumer requirements. Packing material properties include its ease of use, rigidity or flexibility, transparency or opaqueness, user-friendliness, and eco-friendliness. Package should be designed based on product characteristics such as level of perishability; product weight; suitable package dimensions and package styles and desired shelf life. Parameters that are to be considered while designing a polymer-based package for fresh fruits and vegetables are polymer engineering properties such as permeability, Target OTR (Oxygen Transmission Rate), Target WVTR, and tension puncture. Marketing and consumer requirements include protection, presentation and insulation of the package. Presentability of packaging is important to boost the sales appeal of the product. High quality graphics, multi-color printing, distinctive lettering, and logos are being widely used for creating visually appealing packaging designs to boost the sales appeal and visual merchandising at the retail end.

Besides these, converting and filling machine requirements must also be taken into consideration.

Two basic types of designs used for fruit and vegetable packaging are bulk packages and consumer packages. Major bulk packaging designs are CFB boxes of various sizes and shapes and jute-based and polymer-based bags of various sizes and shapes. Consumer packaging designs include trays, punnets, pouches, carry bags etc. Bulk packages are designed keeping producers, processors and wholesale buyers in mind while consumer packages are designed for retailers and consumers. Mainly rigid and semi-rigid packaging options are used for bulk packaging while all three types of packaging methods (flexible packaging; semi-rigid packaging and rigid packaging) are popular for consumer packaging of fresh produce.

7.1   *Bulk Packaging*

Poplar bulk packaging designs that are used for fruit and vegetable packing are CFB boxes; jute bags, lenobags and net bags. These bulk packages are ecofriendly, recyclable and reusable. They provide good aeration to the products and thus reduce product decay and wastage. Jute bags and lenobags are recommended for bulk packaging of highly perishable horticultural produce. A stack load test carried at the Indian Institute of Packaging[1] as a part of fresh produce packaging experiment revealed that at the end of 40 days, only 12% spoilage occurred in jute bags and 8% spoilage occurred in leno bags.

---

[1] Report on Packaging of Fresh fruits and Vegetables for Exports (2005). Indian Institute of Packaging, Mumbai

7.1.1 CFB (Corrugated Fiber Board Box)

CFB bulk packages are designed considering the net weight of products; nature of the fruits and vegetables to be packed and mode of transportation. CFB boxes with molded pulp trays and CFB boxes with slotted partitions are used for bulk transport of fresh produce. Two types of CFB boxes available in the market are RSC type CFBs and telescopic type CFBs.

7.1.1.1. CFB Boxes with Molded Pulp Trays

Each CFB box contains several pulp trays with cavities to hold individual produce. This type of packaging is commonly used for oranges.

7.1.1.2. CFB Boxes with Slotted Partitions

Slotted partitions of CFB are placed inside a big box and individual fresh commodities are placed in each slot. After filling the slots, individual layers are covered by CFB plates. Finally each box is sealed and secured using tapes.

7.1.2 Jute Bags

Jute bags are made from jute fibers, the strongest vegetable fiber available today. Bags are made in different sizes of 25 Kg/50 Kg/100 Kg capacities and can be made into any shapes by stitching. Jute hessian bags are considered to be the best packaging solutions for the bulk packaging of potatoes and onions. Major advantages of jute bags as a packaging option include its biodegradability, ease of handling, suitability to manual loading and unloading process, and cheap availability. Jute bags are economically viable and easy to dispose off. On the negative side, jute bags are highly prone to insect-pest damage and are not moisture resistant. Sometimes wetness of the jute bags may cause fungal and mold infestation of the packed products, and thus quality of the products may be deteriorated.

7.1.3 Polymer-based Bulk Packages

Major polymer-based bulk packages used for fruit and vegetable packing are leno bags, raschel bags and net bags. These bags are available in different sizes of 25 Kg/50 Kg/100 Kg capacities. Like jute bags they provide good aeration to the packed commodities but are much stronger than jute bags in comparison. These polymer-based bulk packages are recyclable, reusable, light in weight, resistant to moisture, resistant to insect damage, resistant to fungal and mold infection, and provide good product visibility. They eliminate pack condensation and therefore prevent spoilage and wastage of the produce. After the filling process, bags are sealed and secured by stitching.

7.1.4 Bulk Packaging Design for Air and Sea Transport

According to the ISO standards, two standard pallet sizes that can be used for the air transport of fruits and vegetables are 1200 x 1000 mm and 1200 x 800 mm while standard internal dimensions of a 20 ft refrigerated container for the transport by sea should be 5364 x 2255 x 2255 mm.

7.2 Consumer Packaging

Retail market for fresh fruits and vegetables is flooded with various packaging solutions and a wide range of paper-based, polymer-based and jute fiber-based consumer packages are available for consumers to choose from. Among these options, polymer-based consumer packaging has created a niche in the market because of its flexibility to suit to ever-changing consumer preferences. While paper-based and jute-based packages are available in very limited designs like paper bags, paper boxes, jute bags etc, polymer-based consumer packages are available in a wide range of sizes, shapes and designs. Popular polymer-based packaging designs are flexible pouches, rigid trays, semi-rigid trays, punnets, tubular nets and carry bags.

7.2.1 Flexible Pouches

Flexible pouches are multiple laminates provided with an outer layer, middle layer and inner layer for complete product protection. Flexible pouches can be made from low density polyethylene (LDPE), polyester and polypropylene (PP). Flexible pouches are light in weight, transparent, flexible (adapts to available space), highly permeable, easy to open and heat sealable (heat penetration is high). Pouches are also provided with ventilation holes so that air is circulated among the packed products and thus rapid product deterioration is prevented. Other features of flexible pouches include its good printability and strength. Besides, flexible pouches are user-friendly and highly preferred by the consumers because of the good product visibility offered by the pouches. On the negative side, these pouches cannot be filled easily and are vulnerable to tearing. Large pouches are difficult to handle and they provide no support to fragile contents. Amcor Flexibles is a world leader in supplying flexible packaging solutions for the fresh produce industry. Sealed Air Corporation is also a leading supplier of various flexible packaging options for the retail sector. Flexible pouches and trays are used for packing salad mixes, fresh-cut vegetables, and fresh fruits.

7.2.2 Punnets

Punnets are best rigid packaging options available for fruit and vegetable packaging. Punnets are made from Polyethylene terephthalate (PET) and hence provide high clarity packaging offering an all-round view of the merchandise and thus aiding consumer choice.

Punnets are highly reusable and recyclable and have excellent lightweight properties. Punnets are provided with a locking system (lids) to secure the packed products and therefore very easy to handle. Punnets also have vent holes to allow air to circulate and fruit to breathe. Punnets are available in a wide variety of formats to suit the consumer and specific-produce requirements. LINPAC Inc is a major industry leader in supplying semi-rigid and rigid packaging solutions such as crystal clear punnets.

7.2.3 EPS Trays

It is a semi-rigid packaging option. EPS trays are made from expanded polystyrene (EPS) and are cost-effective. These are best low-carbon packaging options available today. Standard EPS trays are available in a range of different sizes and colors and offer excellent product protection, presentation and insulation. They have excellent lightweight properties and are suitable for automatic, semi-automatic and manual overwrap applications. EPS trays are recyclable, reusable and extremely user-friendly. EPS trays are generally used for the packaging of vegetables and fresh cut produce. Products are arranged on an EPS tray and then stretch wrapped with a semi-permeable cling film. Industry leaders in supplying semi-rigid and rigid packaging solutions such as EPS trays are AMCOR, Sealed Air Corporation and Bosche Packaging.

7.2.4 Cling Films

Cling films are used for wrapping medium and large sized vegetables and fruits in order to protect them from rapid spoilage. A cling film is a thin semi-permeable transparent film which is capable of clinging to the smooth surfaces of fruits and vegetables. Once a fruit or vegetable is wrapped with a cling film, it produces a modified atmosphere within the packing and thus helps the product regulate its respiration and associated physiological processes for shelf life extension.

7.2.5 Raschel Bags

Raschel bags of various sizes and shapes are made from high density polyethylene (HDPE) using appropriate knitting machines. These seamless bags are soft yet strong and are ideal for packing large and medium sized fruits and vegetables such as potato, onion, carrot, cabbage, cucumber, etc. Raschel bags are available in different colors (red, yellow, orange, green, and violet) according to consumer requirements. There are two types of raschel bags: single bags and bags on roll. Single raschel bags come in two types, one is with drawstring and the other is without drawstring. Raschel bags are ecofriendly, portable, ventilate and highly economical.

7.2.6 Netlon Bags

Netlon bags are made from nylon. These bags are stretchable to accommodate all sizes and shapes of fresh produce and therefore it is the best option for the packing fruits and vegetables of different sizes and shapes. Netlon bags are provided with a locking clip cum holding arrangement for easy consumer handling and can accommodate products up to 400 mm diameter.

7.2.7 Leno Bags

Lenobags are normally made from polypropylene and are widely used for packing onions, potatoes, corns, garlic, cabbages, citrus fruits and flowers. These bags are hemmed top with or without drawstring and the bottom closure is done by chain stitch or over-lock stitch. Leno bags being permeable allow the air to pass through the bags which help to maintain the product's 'fresh-like' quality. Products can be kept safe and fresh for long durations in leno bags. High permeability of the leno bags also helps in saving energy cost in cold storage systems. Being low weight and highly cost effective, leno bags are considered as a superior packaging alternative to other materials. Leno bags exhibit a high level of flexibility and efficiency. Lenobags have excellent reusability and washability characteristics and are available in a wide range of attractive colors. These bags offer good product visibility and hence products inside the bags can be seen very clearly by the consumer for quality check.

7.2.8 EVA Tubular Nets

Tubular nets are widely used for small retail ready packaging of fruits and vegetables. Polyethylene (PE) and Ethylene Vinyl Acetate (EVA) are normally used for manufacturing tubular nets. Tubular nets made from EVA are suitable for automated line-packaging of garlic and onions.

7.2.9 Knitted PE Bags

Knitted Poly Ethylene Bags are available in various types such as mono PE bags; circular PE bags and extruded bags. Mono bags are knitted of strong monofilament polyethylene and are stitched at sides and top with drawstring which makes packaging very easy and convenient. Mono bags are durable and hence highly suitable for heavy produce packaging. Circular bags are made of polyethylene tubular net. These bags are sewn at top and bottom and thus ensure safe product packaging. Circular bags are available in different types of drawstring, and in different colors. Polyethylene extruded bags are made from extruded tubular net and are widely used for small retail packaging. Extruded bags are available in different meshes, colors and labels to meet various product-specific requirements. Standard 4-strand mesh bag is an ideal solution for packaging of soft-skin fruits such as grapes and tomatoes.

8. Packaging Standards

Different types of packaging styles are adopted for fresh horticultural produce for various purposes. For example, a temporary basic packing is used for product protection when freshly harvested produce is packed at the field level. After that, more advanced packaging options are adopted at the precooling and packing facilities. Packaging style is selected based on the product nature, transit distance and storage duration. Codex Alimentarius standards are generally followed for the packaging of fresh fruits and vegetables at various levels. Major highlights of these standards are given in Table 4.

Table 4: CODEX Packaging Standards for Fruits and Vegetables

Strength of the Packaging	According to the Codex Standards a packaging must withstand rough handling during loading and unloading; compression from the overhead weight of other containers; impact and vibration during transportation; and high humidity during precooling, transit, and storage
Selection of the Packaging Materials	Selection of the packaging materials should be based on product requirements; packing methodology; transit and storage requirements; buyer specifications and economic considerations
Suitable Packaging Options	Package options that are allowed include ➤ paper-based materials such as paper bags, paperboard or fiberboard bins, CFB boxes (glued, stapled, interlocking), paper lugs, molded pulp trays, flats, CFB dividers or partitions, and slipsheets ➤ wood-based materials such as wooden bins, wood crates (wirebound,

	nailed), bamboo baskets, wooden trays, wooden pallets
	➤ polymer-based materials such as polymer sleeves, plastic liners, plastic bins, boxes, trays, bags (mesh, solid), containers, film wraps, and pads
Standards for Shipping	According to the Codex standards, best shipping containers are bins, boxes, crates, trays, lugs, baskets, and bags. Packaging styles may include one-piece slotted box with glued, stapled, or self-locking flaps; two-piece half slotted box with a cover; two-piece half slotted box with a full telescoping cover, providing strong walls and corners; three-piece Bliss-style box featuring stapled or glued ends providing strong corners; one-piece box with full telescoping cover; two-piece, die -cut style box with full telescoping cover; and/or one-piece box with wire or fiberboard tabs or hardboard end inserts and plastic end caps, providing stacking strength and alignment
Packaging Options for Ice-packing	Best packaging options for ice-packing include fiberboard boxes that are wax-impregnated or coated with water resistant material
Additional Protection of Produce	Pads, wraps, and sleeves may be used to reduce bruising. Pads may also be used to provide moisture for produce (eg: asparagus); provide chemical treatment to reduce decay (eg: sulphur dioxide pads for grapes); and absorb ethylene (eg: potassium permanganate pads in boxes of bananas and flowers)
Use of Additional Materials	Plastic film liners or bags may be used to retain moisture. Perforated plastic may be used to allow exchange of gases and avoid excessive humidity. Solid plastic may be used to seal the produce and provide for modified atmosphere by reducing the amount of oxygen available for respiration and ripening (eg: bananas, strawberries, tomatoes and citrus fruits)
Packaging Options for Field Packing	Corrugated fiberboard boxes, plastic crates or wood crates may be used during harvesting. Produce may also be wrapped

Packing during Precooling	Shed packing - produce is brought from the field to the packing shed in bulk in field crates, bins, or trucks and after precooling, packed indoors or under cover at a central location. The produce may be precooled either before or after they are placed in shipping containers according to the nature of the produce
Types of Packs	Types of packs include: volume fill (produce is placed by hand or machine into the container until the desired capacity, weight or count is reached); tray or cell pack (produce is placed in molded trays or cells which provide separation and reduced bruising; place pack (produce is carefully placed in the container. This provides reduced bruising and a pleasing appearance); consumer pack or prepack (relatively small amounts of produce are packaged, weighted, and labeled for retail sale); film or shrink wrap (each fruit or vegetable is individually wrapped and sealed in film to reduce moisture loss and decay. The film may be treated with authorized fungicides or other chemicals); modified atmosphere (individual consumer packs, shipping containers, or pallet loads of containers are sealed with plastic film or bags. The oxygen level is reduced and the carbon dioxide level is increased. This reduces produce respiration and slows the ripening process)
Filling of the Shipping Container	Large containers which are very wide and weight more than 23 kg (50 lb) encourage rougher handling, produce damage, and container failure. Overfilling causes produce bruising and excessive bulging of the sides of the container, which leads to decreased compression strength and container failure. Under-filling also causes produce damage. The produce is bruised as it moves around inside the shipping container during transport and handling.
Standardization of Containers	Container standardization is desirable due to large number of different container sizes in use. Advantages of standardization is as follows: ➢ Standardized containers utilize the maximum surface of the pallet with no overhang and little underhang

	➢ Standardization provide unit loads and stable mixed pallet loads
	➢ Standardization reduces transportation and marketing costs as unit loads provide for reduced handling of individual shipping containers; less damage to the containers and the produce inside; faster loading and unloading of transportation equipment; and more efficient distribution centre operations
Examples of Unit/Standardized Loads	Standard wood pallets of 1200 x 1000 mm (48 x 40in), 800 x 1000 mm, 800 x 1200 mm, or 1000 x 1200 mm dimensions Standard slipsheets of 1200 x 1000 mm (48 x 40in), 800 x 1000 mm, 800 x 1200 mm, or 1000 x 1200 mm dimensions
Standard Properties That Can be Included	➢ Vertical interlocking tabs between boxes ➢ Boxes with holes for air circulation, which align when the boxes are stacked squarely on top of one another ➢ Glue between boxes to resist horizontal slipping ➢ Plastic netting around the pallet load of boxes ➢ Fiberboard, plastic, or metal corner boards ➢ Plastic or metal strapping around the corner boards and boxes
Advantages of Wood Pallets as Standard Containers	➢ Wood pallets are strong enough to allow storage under load ➢ Provisions for forklift and pallet jack handling may be provided ➢ Air circulation is not blocked. A sheet of fiberboard with holes for air circulation can be used to distribute air across the pallet ➢ Pallets may be provided with an adequate number of top deck boards to support fiberboard boxes

	➢ Boxes can be arranges neatly so that boxes will not overhang the edges of the pallets
Advantage of Slipsheets as Standard Containers	➢ Slipsheets cost less than pallets
	➢ They eliminate the cost of transporting and returning pallets
	➢ Slipsheets made of fiberboard or plastic are strong enough to be clamped and pulled onto the forklift tines or plate for lifting while fully loaded
	➢ Fiberboard slipsheets may be wax impregnated when used in wet conditions
	➢ Slipsheets used in transportation equipment may be provided with ventilation holes for air circulation under the load

Source: Codex Alimentarius

9. Packaging Technology

A fresh produce is physiologically active and continues to function metabolically even after its harvest. A lot of changes happen in a freshly harvested produce such as changes in respiratory metabolism; compositional changes; changes due to physical injuries and physiological disorders; changes due to ethylene biosynthesis and action; and changes due to moisture loss and pathological breakdown. These changes are responsible for rapid deterioration of product quality. Therefore, it is imperative that a good packaging technology must be employed to slow down or eliminate these changes so that the product's 'fresh-like' quality can be maintained and shelf life can be extended for longer periods. As a general practice, freshly harvested plant products are immediately transported to nearby pack houses for carrying out proper post-harvest operations like pre-cooling (for removing excess field heat from the products); for cleaning and disinfestations processes; and sorting, grading, packing and storage operations. Since fresh products are easily vulnerable to the rigors of material handling, loading and unloading, storage and distribution practices, a proper packaging technology that enables the produce to withstand these pressures, at the same time, also restricts its metabolic processes must be employed. Characteristics of a sound fruit and vegetable packaging technology are protection from mechanical damage; prevention of microbial contamination; temperature regulation; regulation of produce metabolic activities; and prevention of water loss. An illustration of various processes at a packing center is given in figure 1 and figure 2.

Figure 1: Process Flow at a Packing Facility

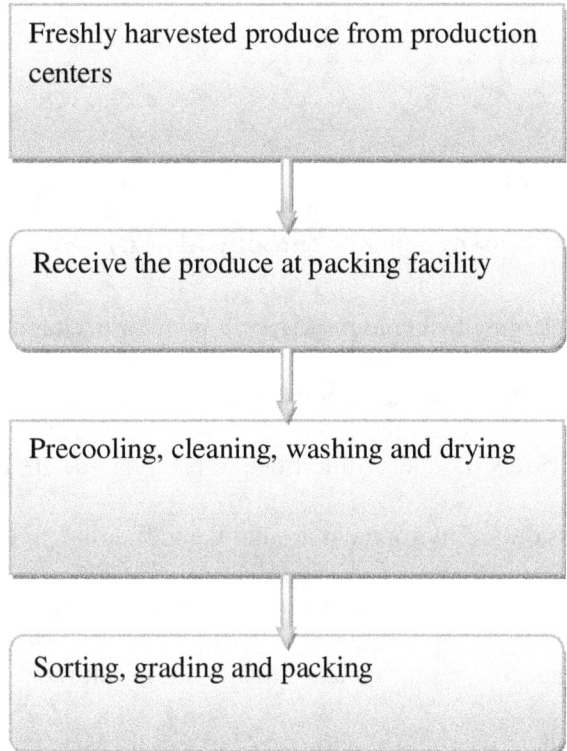

Figure 2: Packing Process

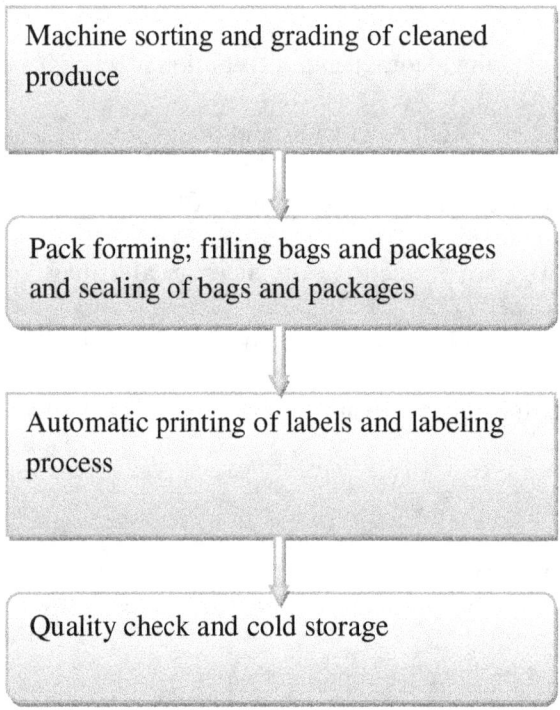

9.1 *Modified Atmosphere Packaging Technology (MAP)*

Modified atmosphere packaging technology (MAP), a popular packaging technology of modern times, is used to extend the shelf-life of perishable fruits and vegetables. In MAP, the environment around the individual products is modified to bring out desired results of shelf-life extension and product safety. Environment around the individual products can be modified either by creating a vacuum in the package (Vacuum Packaging), where there is almost a complete absence of gas or by using special films to allow naturally respiring produce to form its own atmosphere without the addition of external gases. For certain commercial applications, MAP is formed by an optimal blend of gases within a high barrier or permeable package. This process is called 'Gas Flushing'. Generally, oxygen, carbon dioxide and nitrogen are used as MAP gases. Sometimes carbon monoxide and other inert

35

gases are also used however, it is not recommended due to their toxic nature. Modified atmosphere around the individual product restricts the life activities of the product as well as the microbes. MAP is also known as protective atmosphere packaging or reduced oxygen packaging. MAP is considered to be the best packaging solution for packing meats and meat products; egg and poultry; fish and fish products; dairy products; ready to eat meals; bakery products; fruits and vegetables; snack foods; dried foods; liquid foods and beverages and fresh cut fruits and vegetables.

MAP Facts: The best possible quality for a fresh produce exists at the time of its harvest and the harvest quality of this produce cannot be improved but can be maintained by adopting proper MAP practices. The rate at which a product deteriorates depends on certain factors such as the physical structure and properties of the product, the type of microorganisms present and the environment in which the product is kept. Hence adopting proper post harvest management practices is of critical importance for the product shelf life extension under MAP. Therefore factors that affect the product quality must be controlled for shelf life extension of fresh produce.

Advantages of MAP: MAP has now become an essential practice for preserving the fresh quality of the fruits and vegetables. MAP technology restricts the metabolic processes of the fresh produce that are responsible for rapid product deterioration.

MAP reduces ethylene biosynthesis: Ethylene is a ripening hormone present in all fruits and vegetables in varying degrees. Mechanical injuries of fruits and vegetables accelerate ethylene production and therefore accelerate rapid ripening and product deterioration process. Some examples of ethylene effects include, yellowing and abscission of leaves in broccoli, cabbage, Chinese cabbage, and cauliflower, accelerated softening of cucumbers, softening and off-flavor in watermelons, discoloration and off-flavor in sweet potatoes, sprouting of potatoes, increased ripening and softening of mature green tomatoes, increased toughness in turnips and asparagus and bitterness in carrots and parsnips. In MAP, factors responsible for accelerating ethylene synthesis is carefully considered and controlled by controlling temperature and by minimizing microbial activity through high standards of product hygiene.

MAP regulates product temperature: Temperature also has an effect on the shelf life of the fresh produce. Exposure of fresh produce to extremes of heat or cold may cause serious physiological damage to the product which may lead to rapid deterioration. It may also cause product sweating which in turn accelerates product deterioration. Exposure of fresh produce to extreme temperatures may cause heat injury, symptoms of which include bleaching, surface burning, scalding, uneven ripening, excessive softening and desiccation (water loss). In MAP, it is a mandatory practice to maintain an effective cool chain management system and therefore the adverse effects on product quality due to temperature variations are highly minimized.

MAP prevents moisture loss from products: Another major factor that adversely affects the product quality is its moisture content. Moisture loss from fresh produce accelerates its deterioration by shriveling or shrinking the products and by reducing the fresh weight of the product. Factors that affect moisture loss from fresh produce are relative humidity; temperature of the product; environment in which it is kept and air velocity. In MAP, moisture content of the products is controlled by selecting and using product-specific polymers as packing materials.

MAP ensures product safety and quality: MAP inhibits microbial growth and controls all post harvest factors that result in product deterioration. Therefore shelf life of fresh produce can be extended up to 50-400% using MAP.

MAP reduces product losses during long transit time: MAP extends product transit time during transport and therefore product losses are reduced to a minimum.

Disadvantages of MAP: MAP is not suitable for all products as product-specific characteristics play a crucial role in the performance of a modified atmosphere packaging. Since temperature control is of critical importance in MAP, a broken cool chain may result in poor performance of packaging. Major technological challenges faced by MAP sector are in the field of developing product specific gas formulations and compositions. Developing special packing materials and packing machines for MAP also face lots of challenges. MAP industry is also struggling with the development of MAP packages that maintain oxygen partial pressures within tolerance levels as packages undergo changes in temperature and humidity. Challenges also remain in avoiding adverse physiological responses to modified atmospheres. Overall, MAP is an expensive packaging technology which involves additional packaging costs.

9.1.1 Types of MAP

Various types of Modified Atmosphere Packaging are Modified Humidity Packaging; Hypobaric Packaging; Vacuum Packaging; Gas Packaging/Gas Flushing and Equilibrium Modified Atmosphere Packaging.

9.1.1.1. Equilibrium Modified Atmosphere Packaging

When a fresh produce is packed within a MAP packing film, its respiration reduces and as a result, oxygen concentration is reduced and correspondingly carbon dioxide concentration is increased. Produce respiration further lowers until an equilibrium state is reached between oxygen and carbon dioxide concentration. This state is called an

Equilibrium Modified Atmosphere. At equilibrium modified atmosphere, oxygen uptake and carbon dioxide release by the product equals the oxygen and carbon dioxide permeability of the bag. At this point, ethylene synthesis is reduced resulting in delay in product senescence and chlorophyll degradation, therefore product yellowing is reduced. Microbial growth is also reduced and therefore rapid product deterioration is delayed. Break down of sugars and other nutrients is reduced and as a result product quality is maintained. In a nutshell, MAP can increase shelf life; slow microbial growth; maintain nutritional value of foods and slow down browning or discoloration of the produce but it cannot stop microbial growth and improve the quality of the packed commodity.

9.1.1.2. Modified Humidity Packaging

When a fresh produce is sealed in a MAP bag, its transpiration rate increases and as a result, relative humidity within the bag increases up to 90-95 percent. This modified humidity (MH) is beneficial for maintaining product quality because MH reduces dehydration of the product and thereby prevents its weight loss. However, there are certain risks associated with modified humidity. If product-specific MAP film is not used, MH may result in increased microbial activity and therefore product deterioration may be accelerated. Not only this, but MH may also enhance physiological activities of the packed produce (for example, sprouting of buds, regrowth of leaves etc). There may also have higher incidence of physiological disorders such as browning or product discoloration. Such risks associated with MHP can be completely eliminated by using custom-engineered MAP film for each specific produce. Generally, use of a MAP film having high WVTR (water vapor transmission rate) and low OTR (oxygen transmission rate) is the best solution for eliminating the risks associated with modified humidity. MAP film having high WVTR releases excess moisture present inside the package into the surrounding atmosphere and its low OTR can be manipulated to maintain optimum gas composition inside the package.

9.1.1.2.1. *Blending of Modified Atmosphere and Modified Humidity*

Packaging industry is now experimenting with a blend of Modified Atmosphere and Modified Humidity technologies for optimizing packaging solutions for the fresh produce. While non-perforated films are generally used for atmosphere modifications, perforated films may provide the additional option of controlling moisture loss as well. In this combined technology, product quality is maintained and product shelf life is prolonged up to the maximum possible extent due to the combined effects of Modified Atmosphere (high CO_2 and low O_2); Modified Humidity (90-95 % RH) and condensation control (removal of excess moisture).

9.1.1.3. Vacuum Packaging

Vacuum packaging makes use of a range of low or non permeable films (barrier films) and containers for MAP. When a horticultural commodity is placed into the pack, the air is removed and the pack is hermetically sealed under a vacuum with the help of a vacuum packaging machine. No other gases are added to replace the space created by the air removed. However, certain advanced vacuum packaging machines are capable of 'Gas Flushing' i.e. a flush of gases are added to replace the space created by the air removed. These advanced vacuum packaging machines are capable of finishing the whole series of procedures like vacuum extraction, gas flushing, sealing, printing and cooling. The products, thus packed have comparatively longer shelf life than the products packed with simple vacuum extraction technology.

9.1.1.4　　　　Gas Flushing

When MAP gases are flushed in definite proportions into a package to modify the atmosphere around the products, it is called gas flushing. MAP gases such as oxygen, carbon dioxide and nitrogen are classed as food additives under two Acts, the Directive of Food Additives (89/107/EEC) and the Directive of the use of food additives other than colors or sweeteners (95/2/EC). Oxygen is used to preserve respiration, growth rate and composition of flora while carbon dioxide is capable of inhibiting microbial activity if added in optimal concentrations i.e. 20-30%.

9.1.2　MAP Technology from Industry Leaders

9.1.2.1　　　　FreshSpan from DuPont

FreshSpan, DuPont's Modified Atmosphere Corrugated Packaging System consists of a breathable plastic membrane in the liner of the walls of a corrugated paperboard box, which can be hermetically sealed after the introduction of products. During the storage of produce in the box, carbon dioxide concentration increases and oxygen concentration decreases until an equilibrium modified atmosphere is established and a high humidity is maintained. This system inhibits the expression of hydrolytic enzymes associated with fruit ripening and vegetable decay, maintains cellular structure and reduces microbial viability and growth, and thereby inhibits spoilage. In the FreshSpan packaging system, a ratio of 1:1 carbon dioxide and oxygen can occur (as against the standard ratio of 3:1) during the transfer through the membrane. FreshSpan system requires product cooling prior to sealing and temperature management during storage and distribution. The FreshSpan system is product, country and application specific. A performance study of FreshSpan Modified Atmosphere Packaging System (which is recyclable) is given in Table 5.

Table 5: Performance of FreshSpan MAP

Product	Performance with FreshSpan MAP	Benefits
Fresh Cut Asparagus	Storage for 28 days at 1 - 2°C, followed by 2 - 5 days at 8°C	Produce freshness maintained with zero weight loss; package is tamper evident, 100% recyclable
Broccoli	Shelf life of up to 28 days at 0 -3°C, followed by opening and storage for a further 3 - 4 days at refrigerated temperatures	Produce freshness maintained with zero weight loss; package is tamper evident, 100% recyclable
Cauliflower	Shelf life of up to 28 - 35 days at 0 - 3°C, followed by 4 - 5 days at refrigeration temperatures (8°C)	Produce freshness maintained with reduced weight loss
Avocados	HASS variety can be kept up to 21 days at 4 - 7°C. Green avocados can be kept up to 14 weeks at 4 - 7°C, then stored at refrigeration temperatures, 8 - 10°C, or at room temperature (20°C) for ripening. These avocados can have a further shelf life of 5 - 7 days after box opening.	Produce freshness maintained with reduced weight loss, and reduced fungal infection
Blueberries	Shelf life of 42 days (variety dependent) at 0 - 2°C; for 5 - 7 days after box opening	Produce freshness maintained with reduced weight loss, and reduced fungal infection
Cherries	Shelf life of 4 - 5 weeks	Produce freshness maintained with reduced weight loss, and reduced fungal infection

Peaches	Shelf life up to 6 weeks at 0 - 2°C, followed by opening and storage at refrigeration temperatures for 7 - 10 days	Produce freshness maintained with reduced weight loss, and reduced fungal infection
Nectarines	Shelf life of 6 weeks at 0 - 2°C, followed by opening and storage at refrigeration temperatures for 5 - 7 days	Produce freshness maintained
Plums	Shelf life of 8 weeks at 0 - 2°C, followed by opening and storage at refrigeration temperatures for 5 - 7 days	Produce freshness maintained
Roses	Shelf life of 18 days	For roses shipped by sea, shelf life is extended to 21 days

Source: DuPont official website

9.1.2.2   Maptek Fresh from DuPont

DuPont's **Maptek Fresh,** when appropriately used stabilizes the produce and places it in a state of hibernation, thereby extending product shelf life. It is mainly used for fresh cut produce. A performance study of **Maptek Fresh** Technology is given in Table 6.

Table 6: Performance of Maptek Fresh Technology

Fresh-Cut Produce	Performance with Maptek Fresh System	Benefits
Pineapple Chunks	Shelf life of 34 days is achieved at refrigeration temperatures of 0 - 2°C (32 - 35 °F). Products are packaged in 340 g (12 oz.) packages	Fresh, ready to eat product all natural, no preservatives or syrup
Pineapple Cylinders / Slices	Shelf life of 38 days is achieved at refrigeration temperatures of 0 - 2°C (32 -	Fresh, ready to eat

	35 °F). Cylinders are packaged in 20 oz. packages, with the detached core. Slices are packaged in 14 oz. containers	
Fresh Cut Tomato Wedges	Shelf life of 14 days at a storage temperature of 5 - 7°C (41 - 45 °F)	Cut tomato wedges continue to ripen inside the package
Fresh Cut Kiwi Fruits	Shelf life of 15 -17 days when stored at 5 - 7°C (41 - 45°F)	Product quality maintained with little deterioration and with no juice loss
Fresh Fruit Salad Supreme	Shelf life of 16 days when stored at refrigeration temperatures of 5-7°C (41-45°F). Products are packaged in 340 g (12 oz.) containers	Fresh, ready to eat product
Fresh Fruit Salad	Shelf life attained is 10-12 days; recommended temperature is 1 - 5 °C	All natural product
Tropical/Exotic Fruit Salad	Shelf life of 15 days; recommended storing temperature is 1 - 5 °C	All natural product
Mixed Fruit Salad	10-12-day shelf life; recommended storing temperature is 1 - 5 °C	All natural product
Melon Medley	10-12-day shelf life; recommended storage temperature is 1 - 5 °C	All natural product
Mango Chunks	18-20-day shelf life; recommended storage temperature is 1 - 5 °C	All natural product
Citrus Medley	Recommended storage temperature is 1 - 5 °C, and shelf life attained is 18 days.	All natural product

Source: DuPont official website

44

9.1.2.3   Xtend MA/MH Packaging from StePac

Xtend MA/MH packaging creates a produce-specific modified atmosphere (high carbon dioxide and low oxygen) and maintains the proper humidity level inside the bag, allowing excess moisture to escape into the environment. This system blocks biosynthesis of ethylene, thereby slowing down senescence; maintain the nutritional value of produce; reduce decay; slow yellowing of green tissues by preventing chlorophyll degradation and inhibit discoloration of cut surfaces. Xtend packaging makes use of polymers with different water vapor transmission rates (WVTR) for humidity regulation. Performance of Xtend packaging is not only due to the produce-specific modified atmosphere that develops within the packaging, but also largely to the hydrophilic nature of the films that are used. Performance of various StePac MAP solutions is given in Table 7.

Table 7: Performance of StePac MAP Solutions

StePac Product	Benefits
Bulk Bags for "Fully Mature" Mango	o Store "Fully Mature Mango" (waxed or non-waxed) for up to 35 days at 10ºC (50ºF) o Slows down ripening o Reduces weight loss o Reduces decay o Preserves firmness and smoothness o Slows breakdown o Reduces incidence of jelly seed, lenticel spots and chilling injury

Bulk and Retail Bags for "Ready to Eat" Mango	o Store "ready to eat" mangos for up to 21 days at 10°C (50°F) plus an additional 3 days at 20°C (68°F)
	o Reduces weight loss
	o Slows breakdown
	o Preserves firmness and smoothness
	o Reduces incidence of jelly seed, lenticel spots and chilling injury

Source: www.stepac.com

9.1.2.4   NeoSteam MAP Technology from Mondi Packaging

NeoSteam MAP technology uses MAP film along with a steam-cooking valve for steam cooking of the packed produce. Mondi Packaging's NeoSteam Retortable Stand-up Pouch is a classic example of this technology. MAP film creates a modified atmosphere within the package while the integrated valve opens automatically during the steaming process for the pressure to evaporate. Produce can be cooked instantly in a microwave thus making it convenient for consumers. This system ensures a product shelf life of 6 - 12 months without refrigerating.

9.1.2.5   Refresh MAP Technology from Convex Plastics

Convex Plastics has developed 'refresh MAP technology' for minimally-processed fresh products such as fresh salads, fresh cut vegetables, peeled carrots and mix salads. Refresh films manipulate oxygen and carbon dioxide concentrations within the packages in order to achieve an equilibrium modified atmosphere. Refresh MAP technology comes with anti-fog materials and hence provides high clarity packaging.

9.1.2.6   Cryovac Oxygen-Absorbing Sachet

Sealed Air Corporation has developed a unique MAP technology that contains oxygen absorbing materials (oxygen scavengers) for removing the excess oxygen produced within the sealed packages. These oxygen-absorbing sachets can be used either alone or in combination with other MAP technologies such as gas flushing. One of the major advantages offered by this technology is its ability to protect the packed produce from aerobic spoilage.

9.2   *Controlled Atmosphere Packaging Technology (CAP)*

Controlled Atmosphere Packaging (CAP) and MAP are often used interchangeably to refer to the use of gas mixtures inside the packages though there are certain differences between these two packaging technologies. CAP system CONTROLS the atmosphere around the packed products through the use of external machines for injecting gaseous mixtures in definite proportions while MAP system only MODIFIES the atmosphere around each individual product. CAP system is mostly used for the bulk packing of highly perishable fruits and vegetables while MAP is applicable to each individual product also. CAP is often used with additives such as ethylene absorbing or scavenging products, oxygen scavengers, sulphur pads, and carbon dioxide emitters for shelf-life extension.

Recent innovations in controlled atmosphere technology include use of additives such as ethylene absorbing or scavenging products and oxygen scavengers; use of membrane systems and sieve beds for creating nitrogen-rich atmosphere; controlled atmosphere with low oxygen concentrations (o.7 to 1.5 % o_2); ethylene free controlled atmosphere and dynamic controlled atmospheres (where oxygen and carbon dioxide concentrations around the products are modified through monitoring of produce quality attributes such as ethanol concentration and chlorophyll fluorescence). Major commercial applications of Controlled Atmosphere Technology are now limited to using controlled atmospheres for storing and transporting avocadoes, bananas, apples, pears, kiwifruits and certain other fruits and vegetables without chilling injury. It has been established through various experiments that CA is the best option for the shelf life extension of the ripe bananas and controlled atmosphere transport may be used for long distance transporting of nuts and dry fruits also without compromising on its quality. The potential of CA applications is unlimited and it may be commercially used for quarantine treatment as well as long term storage of several fruits and vegetables. Performance of various fruits and vegetables under controlled atmospheres is given Table 8

Table 8: Performance of Fruits and Vegetables under CAP

Product	Storage life (in months) under Controlled Atmosphere
Nuts and Dry fruits - almonds, cashew nuts, Brazil nuts, walnuts, pecan nuts, pistachio nuts	more than 12
Apples and European pears	6-12
Cabbage, Chinese cabbage, Asian pears, kiwi fruits, persimmon, pomegranate	3-6
Avocado, banana, cherries, grapes, mangoes, olives, onions, peaches, plums and green tomatoes	1-3
Asparagus, broccoli, cranberries, figs, lettuce, muskmelons, papaya, pineapple, strawberries, sweet corn, cut fruits, cut vegetables and cut flowers	Less than a month

9.3   *Modified Interactive Packaging Technology (MIP)*

MIP Technology is employed for the production of 'Breathable Bags' where a packed fresh produce 'mimics' a situation as if it is still attached to the mother plant . MIP technology is invented by the Japanese scientists during the early 1980's when grounded particles of a special grade volcanic rock were incorporated into a mono layer LDPE film. The resultant packaging material when used for fresh produce, created the right atmosphere to achieve similar or better results than the MAP. However, it was found that efficient cool chain management is essential to optimize the results of MIP as in case of MAP. One of the advantages of the MIP technology is that it is able to adapt to the changed scenarios of the product transport and storage because it allows the produce to determine its own rate of gaseous exchange. The permeability of the MIP film has been developed to allow" live" produce to increase carbon dioxide levels and reduce oxygen levels in the air space around the produce. The produce senses this change in atmosphere and slows the metabolism and respiration rate accordingly via its own biofeedback mechanism. Therefore the produce is able to develop its own good and sustainable environment for longer shelf life. As a result of MIP technology, there is never a single set of values for oxygen or carbon dioxide levels within the breathable bags. Modified atmosphere packaging usually refers to the practice of modifying/imposing a particular gas composition around a fresh produce via a feed forward mechanism while Modified Interactive Packaging refers to the ability of a fresh to alter its respiration rate and therefore change its surrounding atmosphere via a biofeedback mechanism. That is, the fresh produce itself lowers its metabolism down from a high rate at packing to a much lower, steady rate in storage. Normal PE, PP and PS packaging materials have very low permeability and therefore product metabolism is lowered creating a detrimental anaerobic environment within the package. "Inter-active Packaging" provides higher gas permeability so as to allow greater inflow of oxygen and outflow of carbon dioxide from the package surrounding the produce, thereby allowing the produce enough flexibility to modify its surrounding air so as to survive in a state of reduced metabolic activity. Major differences between MIP and MAP are, MIP technology is more user friendly than the MAP. In MIP, one film can be used to suit most types of fresh produce while product-specific packing material has to be selected in MAP technology. In MIP, packing inventory is substantially reduced with no chance of mixing up or using the wrong film or bag. In MIP, it is NOT necessary to use a cable tie (can puncture bag), instead, an elastic band can be used. MIP films can also be opened and closed without any undue damage to the contents as the internal atmospheres are readily reestablished by the produce once the bag is re sealed

9.4   *Active and Intelligent Packaging Technology*

It is now beyond doubt that certain gases and microbial contamination play critical roles in product quality deterioration and shelf life reduction of the fresh horticultural produce. Major shelf life – reducing gases include oxygen, carbon dioxide and ethylene while major microbial contamination include presence of bacteria (*Salmonella, E. coli etc*), yeast, molds and fungi. Active and intelligent packaging technology makes use of either gas absorbing materials to reduce/eliminate the concentration of these gases or vapor-release mechanisms for controlling microbial growth or sometimes both in order to maintain product quality and also to extend its shelf-life. Active packaging technology incorporates additives (gas absorbing materials) into the packaging film at the time of film manufacturing or within the packaging containers as sachets at the time of packing. These additives come in two types: scavengers/absorbers and emitters. Scavengers absorb excess gases produced within the package while emitters emit the gases required for the shelf life extension of the packed produce. Some of the best examples of scavengers/absorbers used in active packaging technology are oxygen scavengers; carbon dioxide scavengers; moisture absorbers; odor absorbers and ethylene scavengers (removes excess ethylene and suppresses produce respiration). Examples of emitters are carbon dioxide emitters; and ethanol emitters (antimicrobial properties).

9.4.1 PEAKfresh from Convex Plastics

Convex Plastics is a leader in providing active and intelligent packaging technologies for the horticulture sector. Its PEAKfresh active packaging technology is one of its kinds and is best suited for the transport of fresh produce in bulk quantity. PEAKfresh is made from a mineral impregnated film that is capable of slowing down the ripening process. It also absorbs and removes excess ethylene gas produced and maintains high levels of humidity at storage.

9.5 *Packaging Technology for Microwaveable Containers*

Microwaveable packaging materials can be manufactured from Polypropylene because of its strength and excellent light weight properties. PP microwaveable packaging is suitable for both microwaves (can withstand a temperature up to $120^{\circ}C$) and freezers (freezable up to $-18^{\circ}C$, blast freeze is possible up to $-40^{\circ}C$). Associated Packaging Inc is a market leader in providing packaging solutions for microwaveable containers. The company also produces 'Dual Ovenable Containers'.

9.6 *TBG Technology for Retail Ready Packaging*

TBG technology provides high protection for a high-abuse bag with excellent oxygen barrier properties and outstanding retail merchandisability. TBG bags also provide optimum patch protection with minimum material that allows maximum product visibility TBG technology is introduced by the Sealed Air Corporation, a global leader in the packaging industry.

10. Packaging Machines

A wide range of packaging equipments are offered by top industry leaders in the field of packaging like AMCOR, PFM Packaging, Lingwood Services Ltd, Bosche Packaging, Claypack Inc, CSS Packaging Machinery etc. and various packaging equipments are available for performing functions like converting packaging materials from one material to another, forming packages from packing materials, filling the packages, and sealing the packages. Automated packing lines capable of performing forming, filling and sealing together in a single process are also available in the market.

10.1   Vertical Form, Fill and Seal (VFFS) Machines

VFFS machines are vertical packaging machines that perform forming, filling and sealing processes together at one go. A VFFS machine transforms polymer films into bags and then fills and seals the packages. Major features of VFFS machines include its flexibility, small size, compactness, and its self supportive structure. This machine is very user friendly and hence easy to use.

10.1.1 GKS-Easy pack

A global leader in providing VFFS packaging solutions is Lingwood Services Limited which has launched a VFFS packaging machine under the trade name, **GKS-Easy pack** in the market. The GKS-Easy pack is a compact vertical packaging machine suitable for packing various products. It transforms film into a bag by using a specially developed system and the machine also fills and seals the bag using a pneumatic seal system powered by the integrated air unit. The GKS-Easy pack is suitable for packing foods in heat-seal materials such as polypropylene and its capacity is up to 18 bags per minute. The machine is operated either manually or automatically. GKS-Easy pack can be used for the production of standard pillow bags and gusset bags.

10.2 Horizontal Form, Fill and Seal Machines (HFFS)

HFFS machines are horizontal packaging machines that perform forming, filling and sealing processes together at one go. A HFFS machine transforms polymer films into bags and then fills and seals the packages.

10.2.1 SCIROCCO

SCIROCCO is a horizontal flow pack machine developed by the PFM Packaging Inc. It is a high-performing machine terms of both packaging speed and hermetic sealing, other features of this machine include its ease of cleaning and high quality machine components. It is suitable for high speed M.A.P. (Modified Atmosphere Packaging).

10.2.2 PEARL

It is another horizontal Flow Pack Machine introduced from PFM Group. It has full servo horizontal pillow-pack wrapper to obtain pillow-pack style packs from a flat reel of heat/cold sealable wrapping material. It is a high quality innovative machine with a sturdy construction with a carbon steel fabricated frame, and a compact design. It has cutting edge hardware and it is highly reliable.

10.3 Automatic Horizontal Wrapping Machine

Automatic horizontal wrapping machines are primarily designed for producing and wrapping various sizes and shapes of packages. Flexibility, cost effectiveness, outstanding quality and reliability are common for these machines. These machines are used for horizontal packaging using films of heat sealable compound materials such as single Paper/PE, Cellophane/PE, Aluminum foil /PE, BOPP/PE, Nylon/PE etc.

10.3.1 Sprinter Flow Wrapper

It is an automatic horizontal flow wrapping machine from CSS Packaging Machinery.

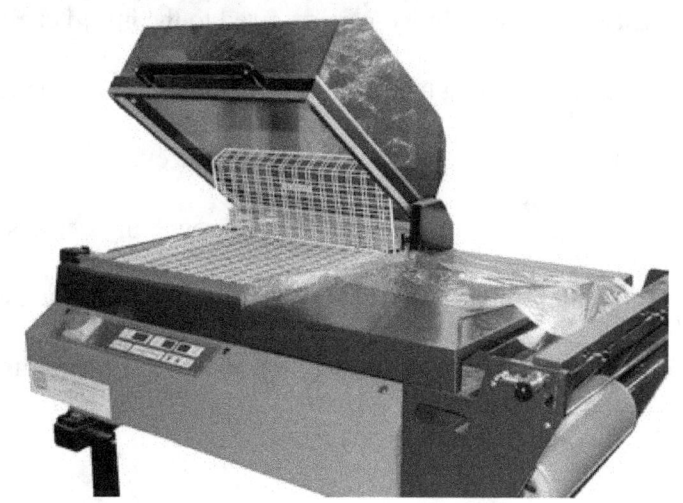

Major features include high speed, high reliability, and robustness of the machine. Sprinter is ideally suited for a wide range of flow wrap packaging applications. It is constructed from stainless steel and anodized aluminum. Sprinter comes with a two meter infeed conveyor. Options are available for fitting features like print registration, overprinter, touch screen and three axis controls. Output capacity is up to 250 packs per minute, dependent upon product size.

10.3.2 Sprinter Junior Top Seal Flow Wrapper

It is another automatic horizontal flow wrapping machine from CSS Packaging Machinery. It is an economically priced machine and is simple to operate. This machine is ideal for products that need to be flow wrapped but cannot be packed on a normal bottom seal machine with pusher infeed. It is constructed from stainless steel and anodized aluminium. Major features of the machine include self centering packaging material mandrel, print registration unit, automatic packaging material edge aligning device, electronic pack length control, a two meter in feed conveyor and a product carry over conveyor through the rotary end crimp.

10.3.3 Pack 201 FV

It is an innovative and versatile horizontal packaging machine developed by Bosch Packaging. Major features of the machine include high reliability, greater adaptability, outstanding accessibility, carriers fixed to the infeed chain to reduce friction and heat development during handling; small support rollers on the deck plates to reduce tensile load on the film, which also avoids film rupture even for heavy products; and easy format changes for packaging various sized products through the stepless adjustable folding box, allow for maximum machine efficiency and improve package quality. Major benefits that the machine offers are delicate product handling; short changeover times; easy maintenance; full stainless steel execution and high flexibility.

10.4 Cling Film Wrapping Machines

These machines are used for wrapping cling films around medium or large sized individual fruits and vegetables, and for stretch wrapping trays. It is a simple machine where two support rollers are provided for the cling film to be used and a stainless steel support bridge attached to it allows the film to pass under the product to be wrapped. A constantly heated, low voltage cutting wire is also provided to cut the film before sealing the bottom of the tray on the hot plate. The hot plate features adjustable temperature settings. Cling films are available in various widths to suit the product to be wrapped.

10.5 Bagging Machines

Bagging machines are used either for bagging or for bag making. An automatic bagging machine normally includes options for weighing, filling and sealing of the bags. CSS Packaging Inc, Claypack Inc and many other packaging industry leaders have already introduced different types of bagging machines in the market.

10.5.1 Compact Bag Maker

It is a simple bag making machine from CSS Packaging Inc. The Compact is a small beam motion machine with a maximum output speed of 35 packs per minute. It is supplied with one set of flat bag change parts and a rear exit pack discharge conveyor. Print registration, over printer, all types of fillers as well as hand feeding can be fitted with the Compact.

10.5.2 BD01 Bag Maker

It is a small, compact bag-making machine from CSS Packaging Inc and has a maximum output speed of 60 packs per minute. It is supplied with one set of flat bag change parts and a rear exit pack discharge conveyor. Print registration, over printer, all types of fillers as well as hand feeding can be fitted with the machine.

10.5.3 Claypack Wicket Bagger

Claypack wicket bagger is an advanced bagging machine from Claypack Inc. It is suitable for high speed weighing of various products like potatoes, onions etc. By means of the two-stage feeding system, the product is evenly distributed over the feed conveyors; hence, overweights are kept to an absolute minimum. By using completely independent speed regulation and two stage main and dribble conveyors, filling speeds can be optimally adjusted and adapted to the product to be packed.

10.6 Strapping Machines

Strapping machines are used for dispensing, tightening and sealing straps (PP or PET) while packing or bundles. Automatic and semiautomatic strapping machines are available in the market. Automatic strapping machine comes with high production speeds and more accuracy rates. Major features include its high efficiency, easiness to operate, high performance, reliability and quick speed.

10.7 Shrink Packaging Machines

Shrink packaging machines are used for packing fresh produce with shrinkable films (PE/PVC). These machines come with the air flow technology that allows air to circulate during the packing operation so that formation of bubbles and bumps in plastic film can be avoided. Major features of a shrinking machine include its efficient heating system to reduce electricity intake and temperature regulation for optimum shrink packing operation.

10.7.1 Shanklin from Sealed Air Corporation

Shanklin is a shrink packaging equipment system introduced by the Sealed Air Corporation in the year 1961. Major features of this machine include its durability, design excellence and quality manufacturing.

10.8 Automatic Case Erector

Automatic case erectors are used for performing packing operations such as carton erecting, vertical forming, bottom flaps folding and bottom tape sealing. These processes are done automatically according to programs. Major features of the machine include its user-friendly interface, high efficiency, high speed, easy size changeover, stability in operation and high reliability.

10.9 Automatic Encasing Machine

Automatic encasing machines are used for placing the lined-up inner packing containers into cartons. These machines are most suitable for dispersed packaging or grouped packaging operations. Both vertical encasing machines and horizontal encasing machines are available in the market.

10.10   Sealing Machines

Sealing machines are used for sealing the packages after completing a packing operation. There are different types of sealing machines available in the market. Carton sealers are used for sealing carton boxes while band sealers are used for sealing thermoplastic bags.

10.10.1 Band Sealers

Band sealers are used for sealing thermoplastic bags of any size, length, and thickness. A variable conveyor belt is provided to adjust the speed in order to accommodate materials varying in thickness. The conveyor belt can be raised or lowered and in some models, even tilted to the desired angle. A band sealer is normally equipped with an ink coding system to record words, dates, and codes with each seal.

10.10.2 Carton Sealers

Carton sealers are used to seal the top, bottom or edges of a carton with tape. These machines are user friendly, reliable, cost-effective, and durable. Carton sealers are available in a variety of configurations to suit various customer demands. Both automatic/pneumatic and semiautomatic carton sealers are available in the market. In semiautomatic carton sealers, the packer has to arrange the uniform-sized cartons together before pushing them into the carton sealer while pneumatic carton sealer automatically detect the sizes of carton boxes and adjust the tape head accordingly. Another version of carton sealer available in the market is Auto flap carton sealer. This machine is equipped with a flap folding mechanism.

10.11 Automated Converting Systems

Automated converting systems are used for converting single or multiple rolls of polymer-based packing materials into customized packing solutions. Cryovac brand of automated converting systems from Sealed Air Corporation is the market leader in this field.

10.12 Feeder Placer Robotic Systems

Feeder placer robotic systems is an advanced packaging technology that combines pick and place robotics with an advanced vision system to place products in a chosen orientation. Average capacity of the machine is 120 individual products a minute. The product arrives on a conveyor belt in bulk where a digital video camera identifies the individual units and a suction pick-up system transfers the product from the belt onto the feeder belt of the flow -wrap machine for wrapping. The robot identifies faulty products and eliminates them before packaging, thus ensuring only high quality products are packed.

10.13 Automatic Packaging Lines

An automatic packaging line automatically finishes the whole packaging process of products right from weighing of the product to its sealing in a package. A packaging line is

capable of performing various packing operations such as sack feeding, sack picking, filling, opening nip, edge folding, sack sealing, empty carton delivering, carton forming, objects-aligning and conveying, robot encasing, auto flap carton sealing, edge sealing, code printing, carton strapping, conveying, robot palletizing, empty pallet delivering, carrying-pallet conveying, vertical strapping, horizontal strapping, wrapping, and sealing. Major features of a packaging line include high speed, high-quality packing, good compatibility and stable performance.

11. Global Suppliers of Packaging Solutions

A comprehensive list of the major global suppliers of packaging materials is given in Table 9.

Table 9: Leading Suppliers of Packaging Solutions

Packaging Supplier	Remarks	Reference Website
AMCOR	A multinational packaging leader	www.amcor.com

Cryovac Sealed Air Corporation	A leading global innovator and manufacturer of a wide range of protective packaging and performance–based materials	www.sealedair.com
Convex Plastics	A company which specializes in the design and manufacture of top quality polyethylene flexible ready-made bags and films, promotional materials and labels	www.convex.com.nz
StePac	Leader in modified atmosphere packaging solutions; two major brands are *Xtend* and *Xsense*	www.stepac.com
CSS Packaging	Leading supplier of packaging machinery in UK	www.cssmachines.co.uk
Claypack	Leading supplier of packaging machinery in UK	www.claypack.com
Associated Packaging Inc.	Supplier of a wide range of packaging solutions which include packaging materials and equipment, shrink and stretch wrap materials and equipment, bagging materials and equipment, technical service, and systems integration	www.associatedpackaging.com
PFM Packaging Inc.	Leading supplier of packaging machinery	www.pfmuk.com
DuPont	A leader in supplying sustainable, eco-friendly packaging solutions and a complete line of additives (antifog, antistat, antiblock and slip agents)	www.dupont.com
Mondi Packaging	Mondi is a leading international paper and packaging group with operations across 31 countries and an average of 31,000 employees	http://www.mondigroup.com

Smurfit Kappa Group	World leader in paper-based packaging	www.smurfitkappa.com
Sharp Interpack	Now known as Sharpak after its acquisition by Groupe Guillin, an international packaging supplier	www.sharpinterpack.com
Styropack Inc.	Leading manufacturer of molded, protective packaging in UK	www.styropack.co.uk
Lingwood Services Ltd	Supplier of Packaging Equipments	www.lingwood.net
NNZ B. V.	A multinational packaging specialist for the agricultural sector	www.nnz.nl
Bosche Packaging	An industry leader in supplying packaging solutions	www.boschpackaging.com
LINPAC	Leading Supplier of punnets, trays etc	www.linpac.com

Bibliography

Wilson Charles L. Ph.D. (2007). *Intelligent And Active Packaging For Fruits And Vegetables* (Trade Cloth ed.). USA: Crc Press.

Bronlund&Robertson. (2006). Modeling of Heat Transfer Through Corrugated Cardboard Packaging. *Proceedings of the International Institute of Refrigeration* (pp. 16-18). Auckland: IRHACE.

IIP. (2005). *Report on Packaging of Fresh Fruits and Vegetables for Exports*. Mumbai: Indian Institute of Packaging.

Kader, A. A. (2002). *Postharvest Technology of Horticultural Crops* (3 ed., Vol. Volume 3311 of Publication (University of California System)). (A. A. Kader, Ed.) California: ANR Publications.

Kirwan, M. J. (2010). *Paper and Paperboard Packaging Technology*. USA: Wiley-Blackwell.

Nicholson&Hertog. (2004). The Effect of Modified Atmospheres on the Rate of Firmness Change of 'Hayward' Kiwi fruit. *Postharvest Biological Technology*, *31* (3), 251-261.

Richard Coles, D. M. (2003). *Food Packaging Technology (Sheffield Packaging Technology)* (1 ed.). USA: Blackwell.

www.ingramcontent.com/pod-product-compliance
Lightning Source LLC
Chambersburg PA
CBHW081606170526
45166CB00009B/2853